CAR CHECKS BEFORE DRIVING

The necessary things to check in your car before you embark on a road trip.

Written by

ANDREW MANICK

CAR CHECKS BEFORE DRIVING

Copyright @ 2021 **ANDREW MANICK**

All Rights Reserved

No part of this book may be reproduced, stored in a retrievable system, or transmitted in any means, recording, photocopying, mechanical, or any other means without the approval of the author and publisher.

Please Note This

Please, be informed that the information contained in this book is meant to spark up the consciousness of checking your vehicle and ensure it's in good condition before embarking on a trip "only".

You will learn a lot from the books. Nevertheless, the "vehicle checks" requirement of your vehicle may defer a little bit from the information contained in this book. Therefore, I will not be held responsible for any issue that emanates from the misuse of this information. Don't replace this information with your car manual, technician, or mechanic advice. This book has the information you need to have a good driving experience. Nevertheless, the necessary precautions should be taken. You will learn a lot from the books.

TABLE OF CONTENTS

TABLE OF CONTENTS.. **VI**

INTRODUCTION **1**

CHAPTER ONE **2**

   YOUR LIFE IS IMPORTANT..2

CHAPTER TWO **4**

   CHECK YOUR LIGHT..4

CHAPTER THREE... **9**

   CHECK YOUR INDICATORS 9

CHAPTER FOUR **16**

   CHECK YOUR WINDSCREEN WIPERS 16

CHAPTER FIVE **20**

   CHECK YOUR MIRRORS 20

CHAPTER SIX.. **24**

   CHECK YOUR CAR HOOTER SOUNDS 24

CHAPTER SEVEN.. **30**

   TEST YOUR CAR BRAKE AND CLUTCH 30

CHAPTER EIGHT 33
CHECK YOUR WHEELS AND TIRES.......................33

CHAPTER NINE 37
CHECK OUT THE TRAFFIC ON YOUR MAP............37

CHAPTER TEN..41
CHECK YOUR FUEL GAUGE..................................41

CHAPTER ELEVEN 45
CHECK YOUR OIL AND COOLANT LEVEL 45

CHAPTER TWELVE 49
CHECK YOUR SEAT BELTS 49

CHAPTER THEITEEN...54
CHECK YOUR SPARE WHEELS AND TIRES 54

CHAPTER FOURTEEN 57
CHECK YOUR STEERING 57

CHAPTER SIXTEEN 63
CHECK FOR LEAKAGES UNDER YOUR CAR............63

CHAPTER SEVENTEEN...69
GET YOUR CAR PAPERS 69

CHAPTER EIGHTEEN **74**

CHECK YOUR BATTERY 74

CONCLUSION **78**

ABOUT THE ATHOR **81**

INTRODUCTION

This book will serve as a guide and reminder to all motorists who want to drive cautiously. It has been ascertained that most people buy fairly used cars without the car manual, while those who buy brand new cars don't spend time reading the manual.

This book is also intended to provide important safe driving tips. To understand these safety tips better and make good use of them, kindly read this book from beginning to the end as every piece of information in this book is vital.

CHAPTER ONE

YOUR LIFE IS IMPORTANT

We often get ready for a journey or trip, whether you're driving long distances or simply faced with increased traffic and intense summer rains in some parts of the world.

While certain factors, such as bad circumstances, are far beyond our control, there is so much we can do as motorists to promote road safety and prevent accidents, beginning with the roadworthiness of our vehicles and the performance of the parts and components in our vehicles.

Every year, millions of people die of road accidents as a result of vehicle malfunctioning. Many people lose important contracts and opportunities that could have given them great riches, but their vehicles broke down on the way and caused delays that deny them the opportunity.

Your life is very important. You need to be live a good life. Read the information in this book and per attention to them.

CHAPTER TWO

CHECK YOUR LIGHT

Make sure all of your lights, not just your headlights, are working. If you're driving in the dark or not, make sure all of your lights are turned on. And if you need to enlist assistance so you can walk around the car and double-check everything. If no one is available, use reflective surfaces such as a window to ensure that all of your lights and functions are functioning properly.

The lighting device on your car or caravan is one of the simplest and most important items to inspect.

While this is something you can check every time you drive, it is particularly important when you're on vacation and heading out onto the highway.

The lights to check before you go on a drive are:

- Car & Caravan Light operation
- Head light lens is clear, clean and not faded

- Head light aiming checked
- Fog lamps are functioning and visible
- Stop lamps are clear and in operation
- Parking light system
- Trailer connections and lighting

Please keep in mind that some vehicles have a more advanced lighting system than what is mentioned here, while others have a less advanced lighting system. It is entirely dependent on the vehicle and its manufacturer. It's important to understand your vehicle and how the lighting systems function.

Check to see if your headlight bulbs are fixed and pointed in the right direction. Bulbs may become misaligned over time and point too far, too low, or to one hand. According to Popular Mechanics, some cars have built-in bubble levelers that help balance the headlights. These are usually found on the headlight unit's sides and top. You'll know the headlights need to be adjusted if the bubbles in the levelers aren't focused.

According to Popular Mechanics, if you need to change the headlights, look for the adjusters and shift them as required. These are usually a bolt or screw on the sides or back of a headlight. You may also refer to your owner's manual for specific instructions on how to change your vehicle's headlights.

If you have any questions when inspecting or cleaning your headlights, seek advice from a mechanic or auto repair store. Know that your headlights aren't the only lighting components on your car that need regular inspection. Check and replace your tail lights, brake lights, and turn signal bulbs on a regular basis.

CHAPTER THREE

CHECK YOUR INDICATORS

Car indicators need a good study to understand what they mean. Please make out time to study your car manual or learn from your car mechanic.

Each warning light is attempting to communicate with you, but you must be familiar with your vehicle's signals in order to decide which warning lights need immediate attention and which can wait until you get home. It may be possible to prevent having your vehicle towed to a repair shop, or it may be necessary.

Dashboard Warning Lights

Dashboard warning lights share similar color-coding characteristics among most manufacturers, even though they are not controlled by the auto industry. Consider it as a traffic signal:

GREEN: Standard vehicle service colors are green. Other colors may vary. It can include items like low windshield washer fluid and high beam indicator.

YELLOW: The check engine light and maintenance lights are usually orange or yellow in color and are of moderate importance. They need to be evaluated as soon as possible.

RED: The color red denotes potential dangers and significant malfunctions that must be addressed right away. The lamps for brake failure, low oil, and engine overheating are usually red. Drivers should pull over to the side of the road and request urgent assistance.

Dashboard Lights Meanings

Some of the most typical dashboard light descriptions are listed below.

Brake warning: In this case, two things could be going on, you might have left the parking brake on (there's usually a different light with a "P" in the circle) or you could have a problem with your brakes, such as a lack of hydrostatic fluid, low brake oil, or extremely damaged brake pads. If your brake pedal feels spongy, take your car to the mechanics for repair as soon as possible.

Engine power restriction: The light will illuminate to show that the power of your engine has been reduced due to failure. No matter how hard you slam on the gas, you won't be able to go faster than a certain speed. To drive normally again, you'll need to have the problem fixed and repaired.

Engine light warning: This means that the pollution system is malfunctioning. If you haven't adjusted your gas cap until you hear it click, the light will usually come on. After you tighten the cap and/or restart the engine, the light can turn off. A steady light indicates the need for service. It may not be an emergency but you need to give attention to it. Stop the car as soon as possible if it's blinking, as this might mean a more severe issue, such as an engine misfire that could destroy the catalytic converter.

Oil pressure warning: Pull over and turn off the engine as soon as possible, since low oil pressure will quickly cause severe engine damage. Either the engine doesn't have enough oil or the oil pump is broken. (Some premium vehicles have a separate light that indicates when the oil level is low.) If your vehicle has an oil level indicator, that's much better.

Engine temperature warning: This sign comes up when your engine is dangerously hot, normally due to a lack of coolant. Switch the car off quickly to prevent engine damage. (If your vehicle has a temperature gauge, turn it off if the needle reaches the red zone.) On a hot engine, never open the radiator cover. The fluid would gush out and it may scald you severely.

Please pay attention to the signals on your car indicators, as they will assist you in determining the state of your vehicle and the appropriate course of action.

Low fuel warning light: When the low fuel indicator light illuminates, put fuel in the tank right away. Engine damage could occur if the engine misfires as a result of an empty tank. When the tank is nearly empty, the low fuel indicator light highlights.

Until the tank is replenished to an internal fuel amount, this light will remain on.

Please refill your tank as soon as possible if you see this indication.

Please keep in mind that depending on the engine and other components in the vehicle, certain cars have more indicators than others. Make an effort to become familiar with the features of your vehicle and attend to them when your attention is required.

CHAPTER FOUR

CHECK YOUR WINDSCREEN WIPERS

To clear the windscreen, your front windscreen wipers must function and be in excellent condition. Please get your rear windscreen wiper fixed as soon as possible if it isn't working.

When it comes to vehicle repair, the wiper and washer system on your windscreen is often ignored, which is not a positive development.

The wiper system's main goal is to keep the windscreen clean enough to provide adequate visibility at all moments. The wiper device must accomplish the following:

- ✓ Dirt, water, and snow are efficiently removed.
- ✓ Operation at temperatures ranging from 240K to 350K.
- ✓ The capacity to pass the stall and snow load test successfully.
- ✓ Wipe period service life is satisfactory.
- ✓ Corrosion resistance to acids, alkalis, and ozone.

Inspect the windscreen for any cracks or chips. If the problem is serious, you'll need a new windscreen. But if you spot it early enough, you may be able to avoid a costly break.

The method by which the blades clean the screen will differ as long as the legally required area of the screen is cleaned. Almost all the wipers are electrically driven.

The wiper on the driver's side must function effectively and efficiently, according to international law. Please remember to respect the laws.

CHAPTER FIVE

CHECK YOUR MIRRORS

Proper placement and use of mirrors inside and outside a vehicle are needed for good visual search habits. The inside mirror can see directly into spaces adjacent to each of the vehicle's rear corners, and the side mirrors can see directly into spaces adjacent to each of the vehicle's rear outside corners using the proper settings.

Mirrors are designed to detect and not to collect information.

The inside mirror: From the driver's seat, adjust the inside mirror so you can see the entire rear window. When using this mirror, you can just switch your eyes, not your head.

If possible, drivers 6 feet or taller will find it helpful to reposition the mirror upside down. This normally lifts the bottom edge of the mirror by 1 to 2 inches, which can help tall drivers significantly reduce blind spots in the front.

Side-view mirrors: Place your head against the left side window and change the driver's side-view mirror so you can see the side of the car on the mirror's right side.

Place the head just above the center console to change the passenger's side-view mirror. Set the mirror to the point that you can only see the car's side on the left side. If your vehicle does not have a mirror-adjustment control, you will need help positioning this mirror properly.

You'll have almost smooth visual communication around your car with these settings, which will help you identify the presence of nearby drivers. When a car passes you in the lane to your left, for example, you'll see it move from the rearview mirror to the left side mirror, and then to your side vision.

Before you get behind the wheel with these new mirror settings, test them out when your car is parked. You can make a parallel park along a street, and then observe passing vehicles through your mirrors and peripheral vision. Before you go out into traffic, this will help you get used to the new settings.

Remember that even the best-placed mirrors won't be able to remove all blind spots. Before attempting any lateral movements, make a final check to the sides to minimize and avoid danger.

CHAPTER SIX

CHECK YOUR CAR HOOTER SOUNDS

Many vehicle horns can be changed if there isn't enough current going through them and the sound isn't loud enough. Horns, unfortunately, cannot be repaired after they are damaged.

Many things can go wrong with a car horn, stopping it from working properly. A typical horn can be examined and troubleshot using a set of basic steps that can be used to troubleshoot most circuits. Only a basic understanding of electricity or a willingness to learn basic troubleshooting procedures is needed.

Instructions

1) If the horn isn't working at all, check the circuit's fuse. Replace the fuse if it has blown, then test the horn again. Proceed to the next move if the fuse is in good working order.

2) Open the vehicle's hood and have an assistant push the horn button on the steering wheel as you listen for a potential poor horn signal. You may not be able to hear the sound because it is too fainting. While the horn is engaged, place your hand on the horn and try to feel a vibration.

Proceed to the next level if you hear sound.

3) Use a regular or any other needed screwdriver, locate the adjustment screw on the horn and adjust it. Continue to the next stage if the horn is still not working.

4) Connect a jumper wire to a better place on the vehicle and have someone push the horn button as you connect the other end of the jumper wire to the horn. If the horn works, make sure the ground connection is secure and that the horn is in good contact with the chassis.

5) Remove the horn from the vehicle and use jumper wires to attach it to the battery. Replace the horn if it

stops working. Proceed to the next level if it works.

6) Reconnect the horn to the circuit and use a voltmeter to check the voltage of the horn, attaching the red probe to the terminal and the black probe to the horn's body. Press the horn button on the steering wheel with the aid of an assistant. Replace the horn if it is receiving voltage. Proceed to the next level if the voltage is ok.

7) Inspect the wire connecting the horn to the relay for continuity. The wire has an open if there is no continuity. Make the required corrections and retest. Proceed to the next level if there is consistency.

8) Make sure the horn relay is in good working order. While an assistant

operates the horn button on the steering wheel, check for voltage at the relay's power-and-control circuit with a voltmeter. Adjust the relay if it isn't functioning properly and test it again. Proceed to the next phase if there is no voltage hitting the relay.

9) Examine the wire that connects the horn relay to the fuse panel. If you discover an open or a short, fix it and retest. Proceed to the next level if the wiring is satisfactory.

10) Have an assistant press the horn button on the steering wheel while you check the wire connecting the relay to the horn button and ground for continuity. If the wire has an open, close it and measure it again.

Replace the horn button if you can't find the open in the wiring.

Please, don't drive a car without a horn that works effectively. Save your life and that of others.

CHAPTER SEVEN

TEST YOUR CAR BRAKE AND CLUTCH

Do you have a spongy brake pedal that travels further than normal, or does your car drift or pulls to one side when driving? If this is the case, have it verified. Allow for a compressive inspection to ensure that these vital components of your vehicle are fully functional and reliable in order to keep you safe on the road.

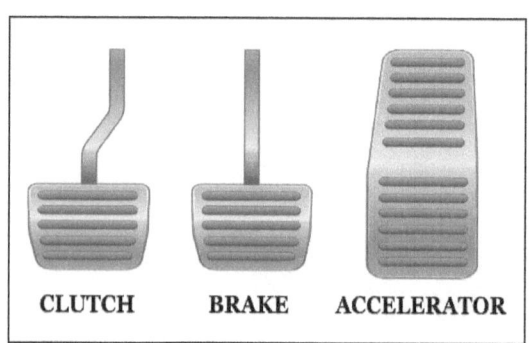

Your vehicle's brake pads should also be tested for cracking and wear. This is a common problem with brake pads that have been overheated or are on their way to being replaced.

While many issues in these areas can seem minor when driving on the highway at high speeds or towing your caravan or trailer, they could become more serious when driving long distances on the highway at high speeds.

There is only one inspection, one result, and one outcome. Before you go on your next road trip, make sure you get this inspected. Keeping you on the road in a healthy and safe manner.

CHAPTER EIGHT

CHECK YOUR WHEELS AND TIRES

The first line of protection for your vehicle against hazardous road conditions is your tires. The tread on your tire aids in channeling water and snow away from your vehicle so you can stop easily and safely. However, as tires age, the tread wears down to the point that they can no longer function properly, placing you and your passengers in danger when you need to stop quickly.

Check that the tread depth indicators (set in the tire groove) are below the tread depth. It's time to change the tire if they're higher than or equal to the tread depth.

Check for signs of tread wear and damage, and make sure your tire pressure is safe, according to your car's owner's manual or mechanic advice.

Replacing faulty and worn tires would go a long way toward improving your vehicle's road safety.

Smooth or damaged tires can cause road accidents by raising your braking distance and preventing you from stopping quickly if necessary. It is often preferable to replace your tires with new ones, as purchasing used tires exposes you to increased risks, with approximately 60% of these tires considered unsafe and dangerous for use on the road.

Remember, you have no idea where the used tire has been, how it has been used, or even whether the grooves have been deepened by an unscrupulous dealer who has duplicated them incorrectly. This is a very unethical activity.

Accidents and other road fatalities are sometimes linked to the use of poor tires, in addition to unsafe cars. If you need a tire for your car, please make sure you get an original tire. Avoid goods that are of poor quality or are counterfeit. They could put your life and the lives of others in danger.

CHAPTER NINE

CHECK OUT THE TRAFFIC ON YOUR MAP

Almost everybody now has access to GPS devices, mobile applications, or in-car technology that can provide directions even to the most obscure locations.

However, knowing how to read a map is useful just in case you need direction to a particular destination. You know, the kind that's written on a piece of paper. It may be difficult to refold the map, because it takes up precious space in your glove compartment. But it is very necessary to have it for easier location. The kind that the gas station attendant uses to draw directions to you. Before the invention of global positioning devices and smartphones, road maps were the only way to get from one place to another.

These maps were, and still are, widely available, often for free, at gas stations, state-sponsored rest areas, and truck stops. Road maps, also known as atlases, are available in a variety of scales, ranging from urban maps that display only a city and its environs to atlases that show major highways and interstates in the state and country. While reading a map is a simple skill, many people, especially young drivers, are unable to navigate effectively with only a paper map.

Before you go through how to use a road map in detail, consider the following scenarios in which knowing how to read a map might be beneficial. Your phone dies, someone steals the GPS that you forgot to put in the console when you were out shopping, your navigational system isn't up to date with new roads or construction zones, and you're in a remote location. Many of these scenarios are very likely, and if you don't know how to read a road map, you'll be looking for directions.

Prepare for heavy traffic levels, which means more time behind the wheel, by keeping snacks and drinks on hand and keeping an eye out for drowsy or distracted drivers. Check the map app's traffic to see which roads are congested.

CHAPTER TEN

CHECK YOUR FUEL GAUGE

The humiliation of being stuck on a road, street, or highway when the tank runs dry is exceeded by the inconvenience that comes with it.

In any case, there's no reason for running out of petrol when there's a fuel gauge right in front of the driver on the dashboard. In addition, there is always a petrol and diesel station within the area, which works within sufficient hours of the day.

Your fuel gauge is crucial in indicating when your fuel tank is running low on fuel and needs to be refilled. You check your fuel gauge to see whether it's full or not, and how much it's loaded when you go for a refill. The fuel gauge can become stuck on zero, leaving you unsure of how much fuel you have left. In the case of a long drive through the countryside, such a scenario can be disastrous.

A fuel gauge is a component that can be found in almost any vehicle. It's the part in charge of sending signals to the instrument cluster's fuel level gauge. It consists of a float, an arm, and a resistor that adjusts in response to the position of the fuel float. The float continues to balance on the top of the substance in the fuel tank. The float and arms change position as the fuel level decreases, and the resistor shifts to power the gauge display within the vehicle.

The two main markings on a fuel gauge are "F" for Full and "E" for Empty. Some also have a "1/2" marking to indicate a half-tank. The fuel gauge essentially only needs to calculate voltage through a variable resistor. However, the resistance of the resistor must differ in direct proportion to the amount of fuel in the tank. A float at one end of a lever arm-and-slider running over a strip of the resistor at the other end of the fuel tank is used to accomplish this aim. The Sending Unit, also known as the S. Unit, is an essential component of a vehicle's fuel tank and functions flawlessly. Please pay close attention to your fuel gauge and fill up your tank if your vehicle's fuel gauge indicates that it is running low.

CHAPTER ELEVEN

CHECK YOUR OIL AND COOLANT LEVEL

Check all the fluids in your car to make sure that they are topped up and capable of performing at their best. It has an effect on the overall health of your car.

Before you embark on a trip with your vehicle, you are advised to check the level of the following oil:

1) The engine oil

2) The Power steering fluid
3) The Transmission and differential fluids
4) The Coolant Anti-freeze
5) The Windscreen washer fluid
6) The Brake fluid.

And other necessary oil that runs the car.

It's crucial to understand how to check car oil before it does more harm to your engine and the many other components that keep your vehicle running smoothly.

If you have a new model car, check your user manual for electronic oil monitors and indicators that will alert you when it's time to change your oil. Traditional dipsticks, which are still an effective way to check your oil in older models, can still be used. If your vehicle has an oil dipstick, you can perform a fast inspection at home using these simple methods:

1) Ensure that the car is parked on level ground with the engine switched off and cold.
2) Locate the dipstick in the hood of the car.
3) Remove the dipstick from the engine and clean any oil from the end.
4) Insert the dipstick back into its tube, then take it out and check the level on both sides.

If the indicator is between the two marks or within a cross-hatched area, the oil is fine. If it isn't, you can add more.

Make sure you get the oil grade recommended in the user manual. Remove the oil filler cap and slowly pour in a small amount of oil until the dipstick shows the correct level.

Please take this advice seriously in order to prevent unfortunate events while driving.

CHAPTER TWELVE

CHECK YOUR SEAT BELTS

According to studies, wearing a seat belt correctly lowers the risk of fatal injuries in the front seat by 50 percent and reduces critical injuries by 55 percent. Rear seat belts reduce fatalities by 75 percent for those in the back.

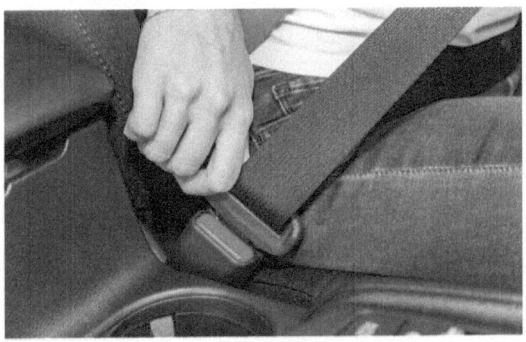

Despite the fact that these advantages are well established, drivers and passengers prefer to ignore them. As a result, more than half of all teen and adult passengers were unrestrained in the 22,441 vehicle collisions. Who is more likely to refuse to wear a seatbelt? Although, men and adults between the ages of 18 and 34 are less likely to wear seat belts, teenagers are the most likely to be seen without them.

CHILDREN EMULATE ADULTS

According to studies, when adults wear seat belts, children in the car are more likely to do so as well. In reality, children wear seat belts 92 percent of the time when adults do. When adults fail to fasten their seat belts, children buckle up just 72% of the time.

WEAR SEAT BELTS CORRECTLY

To avoid accidents, there is a proper way to wear a seat belt. Not behind your back or under your arm, the shoulder strap should cross your chest and away from your neck. The lap belt should cross your hips and sit just below your stomach. The belt should not be too tight to be uncomfortable, neither should it be too loose to allow you to move about in it.

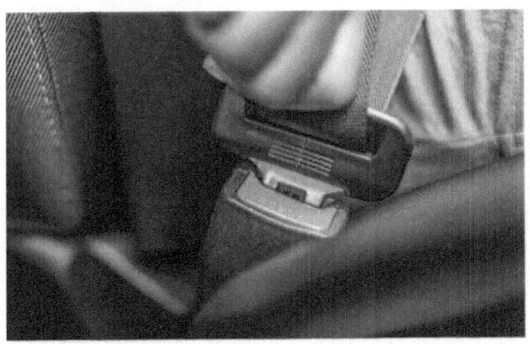

A seat belt is designed to reduce fatal risks, but casualties can occur for a variety of reasons, which includes:

1) It's way too tight on you.
2) It is worn incorrectly.

3) In the case of a crash, the seat belt may malfunction if not properly fastened
4) It may have flaws in the manufacturing process.

However, as related to the injuries that could occur if you were not wearing a seat belt, the injuries from a seat belt are minor. Internal injury to the abdomen, spinal cord, back, spine, or sternum, fracture, dislocation, or internal bleeding are all possible for passengers when there is a fatal accident. It is important you fasten your seatbelt whenever you are in the vehicle to avoid any of this painful conditions.

CAR CHECKS BEFORE DRIVING

AIRBAGS CANNOT REPLACE SEAT BELT

Some drivers feel that if their car's airbags are working, they won't need to use their seat belts in the event of a collision. While airbags provide additional security, if you are not wearing a seat belt, you can slip underneath. Furthermore, airbags are ineffective inside, rear, and rollover crashes. A seat belt, on the other hand, keeps you from falling out the window or being thrown around inside the car.

Please fasten your seat belt while you are in the car, whether or not you are driving. Your life is extremely valuable.

CHAPTER THEITEEN

CHECK YOUR SPARE WHEELS AND TIRES

Spare tire is similar to insurance in that you don't need it until something goes wrong. It's critical to have a high-quality spare tire in your car. Regularly inspecting the wheels and spare tire is also recommended. When you need to use your spare tire in an emergency, it must be in good shape.

The safest choice, if you have a flat tire, is to use the spare tire. A patch kit or inflation device would not be as good as putting the spare tire on. There are occasions when your tire goes flat and all you need is a little more air to repair it. There are other times when air alone will not suffice to repair a flat tire, and a spare tire will be needed to get your vehicle back on the road.

A patch pack is used to make temporary repairs to small holes in the tire. It can also be used if the tire has a moderate amount of air spilled on it. The break is more genuine when the tire becomes fully level or can spill at a rapid rate. If this is the case, repair units are useless because they do not hold air in the tire when driving. Examining the spare tire on a regular basis will help it stay in shape and be ready for use when you need it.

CHAPTER FOURTEEN

CHECK YOUR STEERING

The power steering system applies energy to your steering wheel motions, allowing you to turn the wheels physically. You'd have a lot of trouble steering the car if it didn't have proper functional power steering.

When the power steering fails while the vehicle is in motion, most drivers are surprised at how much force is needed to turn the wheel. This is very dangerous and has resulted in many incidents.

The following are some of the steps you can take to ensure your car's steering is in perfect condition.

1) **Inspect your power steering fluid level**

The function of the entire system is dependent on the power steering fluid. The pump generates pressure, which the fluid will then apply to the piston. The majority of power steering issues occur when a leak has formed and the fluid level is low. When you turn the steering wheel, you will notice that it becomes more difficult to turn and that it makes a squealing noise.

Being stuck on the road without power steering is never enjoyable. We still recommend checking the fluid level before leaving for a long trip and returning. Otherwise, it's a good idea to check the power steering fluid when you have your oil changed.

2) Examine the high and low-pressure hoses

Two hoses are used to carry the power steering fluid. One delivers high-pressure fluid, while the other delivers low-pressure fluid back to the fluid tank. It's not uncommon for these hoses to develop leaks. They should be tested on a regular basis to ensure that they are not rubbing against each other and that the protective coverings are still in place.

3) **Replace the filter when necessary**

The filter should be replaced every year or as soon as it is faulty. Some drivers are completely unaware that a power steering fluid filter exists. This filter should be replaced once a year or as directed by your owner's manual or mechanic. To minimize wear and tear on your pump, keep the fluid clean and free-flowing at all times. You'd be surprised by how many toxins make their way into the fluid.

4) Check the power steering fluid

Changing the filter and holding your fluid clear of contaminants preserves the fluid's consistency and efficiency. You can test the fluid's consistency at home by taking a sample, examining the color, and looking for metal fragments or other dirt and grime. The worst the fluid is, the darker it becomes. The fluid should be flushed and replaced if it is dark and has debris floating around in it.

5) Change the pump if needed

Low fluid levels and dirty fluid may have a negative effect on your power steering pump. It may be a warning that your pump is about to malfunction if you hear a whirring or moaning sound while driving. It's always a good idea to get it checked out by a mechanic. If necessary, replaced it. If a bad pump is not fixed, it can lead to even more costly power steering repairs or damage.

CHAPTER SIXTEEN

CHECK FOR LEAKAGES UNDER YOUR CAR.

Following the fluid level tests under the hood, it's a safe idea to look underneath your car for any leaks from different components and hoses.

It's easy to become alarmed at the first hint of an engine oil leak. Many of us have backed out of a parking spot and been shocked to see a puddle of gasoline, unsure whether or not it is safe to drive our car. It's important to act as soon as you notice a fluid leak, but understanding the causes and effects will help you determine if it's safe to drive your car until it's repaired or whether it requires urgent attention. Hopefully, this short guide will be of assistance.

Please have your car tested by a technician if you see any oil spots or coolant that has a shiny texture that looks like oil on the garage floor.

Identify Which Fluid Is Leaking?

Though brand new cars hardhar leak oil, even low-mileage vehicles will experience some oil seepage. A small amount of oil on the ground isn't really a major issue. When you find a fluid leak, the first thing you can do is figure out what kind of fluid it is.

Engine oil, transmission oil, coolant, power steering fluid, brake fluid, differential oil, and other fluids in your vehicle may all be leaking. Since all of these fluids seem to be the same to the untrained eye, mop up the fluid with a white rag and equate it to the ones in your vehicle. Engine oil can have hints of brown, yellow, or purple, and it darkens with age.

Ascertain The Amount of The Leakage

The most critical thing to consider is how much oil you've lost. If you just see a few drops on the ground, you probably don't need to be concerned. However, it's always a good idea to inspect your engine to see where the leak is coming from. When a leak becomes larger, you can see oil leaking while your car is parked and the engine is running.

The only way to tell if it's safe to drive with an oil leak is to check your oil level regularity of the leakage. On the dashboard of modern vehicles, there is an oil level display that shows when the oil level is low. Some also have a percentage measure for oil life, which is more oriented toward oil shift intervals than oil amounts. You can check the oil level dipstick on older cars to see if it's low. Your oil level should always be in the center, between the minimum and maximum markers. Operating your engine below the minimum oil level is never a good idea, as it can easily cause serious engine damage or even failure.

Repair The Engine Oil Leak

There is no chance of driving with a minor leak as long as the amount of your engine oil does not fall below the minimum. However, for a variety of reasons, we suggest fixing a leak as soon as possible.

A small leak can easily turn into a big one. If this occurs while you are driving, you may lose all of your gasoline, resulting in engine damage.

Oil leaking from your car can cause slick surfaces in your garage, driveway, or parking spot, as well as on the road.

Oil on the floor is a pollutant to the environment. It is poisonous to plants and animals, and it can pollute rivers. It costs money to keep topping off even a small oil leak, particularly with today's expensive engine oils. It's also inconvenient, especially when it comes to peace of mind.

The first warning of an oil spill isn't always a serious issue. It is possible to continue driving in the short term until you have determined that your engine has sufficient gasoline. To avoid being caught off guard by a larger oil spill, it's always best to handle an oil leak as soon as possible.

Consider one of the tested and approved engine oil leak remedies and get back on the road as soon as possible.

CHAPTER SEVENTEEN

GET YOUR CAR PAPERS

Some people keep their car documents in the car at all times, while others carry them while traveling and turn them in as soon as they return home. Whatever route you take, keep in mind that your car information or documents should still be in your vehicle while you're driving.

This will confirm that you are the rightful owner of the vehicle or that you have the proper authority to operate it.

When you're about to hit the road and see the world, double-check that you have all of the required documentation in your vehicle. Certain documents must be carried in your car in every country. If you are stopped by the police, they will expect you to have these papers, and failing to do so could result in a fine of a certain amount of money per document. It might not be a significant sum, but it's still best to avoid any issues if you're stopped by the cops.

There are some documents that you must have in your car and others that we suggest you bring even if they aren't needed.

The following are relevant and obligatory documents to have in your car in most countries.

- A valid driving license.

- Registration certificate: The information on this certificate includes details of the car and its owner.

Please keep in mind that your country's government might need additional documents than those mentioned here. Make sure you have all of your car's correct paperwork, as required by your country's government.

If you lose either of these documents, you must first report the loss or theft to the nearest police station, then schedule an appointment with the necessary authorities in charge of road and traffic in your country to obtain a replacement copy. In the meantime, you can drive, and if a cop asks for your papers, all you have to do is show proof of filing a report. It is often preferable to hold the originals since photocopies of these documents are not accurate unless they are approved.

I don't believe there is any excuse for a driver on the road to not have all of his vehicle documentation. You are not required to bring the original copy with you. Simply make a photocopy and return the original to your house.

The types of documents that should be in the car at all times while driving are often stated by government officials. The document list varies depending on the region. Check the paperwork that the government and law enforcement authorities in your country need to be in your car and make sure you follow all of the rules and regulations. Please act responsibly as a good citizen.

CHAPTER EIGHTEEN

CHECK YOUR BATTERY

How significant is your vehicle's battery? Manufacturers have more electronic components than ever before in today's new vehicle ranges, putting more strain on our cars' battery systems.

The significance of a functioning battery to the operation of your vehicle cannot be underestimated. The car system can malfunction if your battery drops below the required voltage, causing your car to stop or run abnormally.

A car battery usually lasts between 2 and 4 years, so if yours is older than that, it's worth getting it checked before the ride.

You can visually inspect the battery to see if it has any cracks, damage, or acid leaks, or you can have it checked out electronically by one of your mechanics. The battery terminals should be safe and free of corrosion. If your battery fails to pass these visual inspections, it needs to be replaced.

Car batteries can go flat for a variety of reasons, which is why we've compiled this list of helpful hints to help you avoid the worst-case scenario.

A dead battery can be a major pain, particularly during the dark winter months. A flat battery, like many other standard car parts, can be avoided by maintaining it regularly. A dead battery means that your car would not function, no matter how well you looked after it. Cold winter weather has a variety of effects on batteries, slowing down the chemical reactions that take place within. In the worst-case scenario, there might not be enough current to start your car or crank your engine.

CAR CHECKS BEFORE DRIVING

One of the worst ways to start the day is with a dead battery that prevents you from making your journey. As a result, it is advisable to maintain the best possible condition of your battery during the cold, dark winter months. During the winter, dead batteries are the leading cause of breakdown calls.

Please keep this knowledge in mind and make sure your battery is in good working order before driving.

CONCLUSION

After checking all the listed parts and components in your vehicle and confirm they are in good condition, you are good to go.

Please note that because of the constant advancement of technology in this present world, the recommendations listed in this book may be a bit different from the components in your vehicle. All the parts and components listed here may not be in your vehicle. Also, you may have more parts or components in your vehicle than what is listed here. It all depends on the brand and model of your vehicle. Whichever brand or model you use, endeavor to run all the necessary checks before you embark on a trip.

Your life should be as important to you as it is for your loved ones. I know you wouldn't want to have any distressful experience on the road as a result of a car breakdown or critical faults. Please take this information very seriously.

Thanks for reading.

Always check out my other books.

For questions and inquiry, kindly contact me via my email address:

bornlandltd@gmail.com.

ABOUT THE ATHOR

Andrew Manick Is an experienced engineer. He studied Mechanical Engineering and Safety Management. He writes books and other publications on car driving, engineering, personal and public caution, lifeguard awareness, safety, and security. The readers of his books have confirmed the accuracy of the information he passes across the globe for the safety of the people.

If you desire to acquire the right knowledge on safe driving, home protection, and other safety and security awareness, then endeavor to get as many of his books as you can. Stay Safe.

Note

Note

Note

Note

Note

Thank You

For Reading.

www.ingramcontent.com/pod-product-compliance
Lightning Source LLC
Chambersburg PA
CBHW031448210526
45464CB00005B/2369